AMAZON ECHO

Easiest User Guide To Master Amazon Echo Fast!

Welcome to this smart Amazon Echo User guide. From this guide book, you will acquire significant details of Amazon Echo. To maximally benefit from the Amazon Echo, you must understand the details outlined in this book carefully.

I thank you for sacrificing your time to purchase this book. I trust that it will provide the essential details on the use of Amazon Echo. Continue reading!

Table of Contents

Introduction 1
    What the book covers 1
    What is Amazon Echo? 1
    Bringing the smart home to life 2
    What Is Alexa? 2
    How to access the Amazon Echo from the Web 3
    Overview of echo operation 4

Functions of Amazon Echo 5
    Building and managing a smart home 6
    Get the Weather (and Other meaningful Details) 6
    Listen to a publication or a book 7
    Other general uses 7

Features 9
    Voice system 9
    Application updates 9
    Input 9

Apps that boost Amazon Echo functionality 10
    Automatic app 10
    Capital One 10
    Domino's Pizza 10
    Jeopardy! 11
    Quick Events 11
    NBC News 11
    Spelling Bee 11
    The Wayne Investigation 11
    7-Minute Workout 12
    Uber 12
    Play ball 12

Traffic alerts 13

Music playback 13

Inquiries, Conversions, and Trivia.. 13

Shopping and To-Do lists... 14

How to use Amazon Echo 15

Managing the Amazon Echo multiple Amazon accounts 16

Various places where one can use the Amazon Echo 17

In the kitchen... 17

In the bedroom 18

In the living room... 18

Tips to maximally benefit from the Amazon Echo 20

Ways to add a software update to the Amazon Echo........... 21

Amazon echo connectivity 22

controlling your smart things 24

How to set up the Amazon Echo 25

How to customize your Echo experience.............................. 27

The Echo basics you should always remember................... 28

The principal advantages of Echo over its competitors 32

Echo has more data... 32

AWS Backbone 32

Innovations..33

Availability.. 34

Amazon Echo Pros 35

Amazon Echo Cons 36

Suggested Improvements for the Amazon Echo 37

Echo and the future 38

Amazon echo privacy concerns 39

Conclusion 41

Introduction

The present world has technological advancement for the Echo devices available. The Echo devices vary in their potential to provide accuracy in the device functioning. However, the Amazon Echo has proved to be a powerful equipment to handle diverse functions to the users. The consistent upgrading of the device has shown its outstanding performances. I hope by the end of this book you will determine all the capabilities of the Amazon Echo. Interestingly, you will find some uses to match your desires wherever you would be located. Moreover, in most of your daily activities, the Alexa could be a great device to ease your work. You will be able to enjoy distinct musical tastes under different profile on the app. This book gives guidance on the use of Amazon Echo. One must comprehend the running of the Amazon Echo. It has proved to be an outstanding option over other recording devices. Without any doubt, you will choose to use the Amazon Echo services regularly.

What the book covers

The book gives clear outlines of the Amazon Echo. These include descriptions of the Amazon Echo, how to set it up and the ways to use it. Additionally, this book shows various incorporations with the device to ensure high standard outcomes. The benefits and the limitations of the utilization of the Amazon Echo are discussed in this book. Undoubtedly, you will acquire a sophisticated guide on the Amazon Echo. Thus, I'm confident you will have a helpful consciousness to start using the Amazon Echo.

What is Amazon Echo?

Amazon Echo is a smart speaker that is designed by Amazon. The gadget has a 9.25-inch cylindrical speaker together with a seven-piece receiver. It has an incorporated device termed as "Alexa" which helps it to operate. The Echo has a lot of functions making it highly valuable. It is highly supported by the Alexa voice command technology inside it. The properties on the Amazon Echo are currently available on the other devices. These include, Amazon Fire TV, Echo Dot, and Tap. As noted the Amazon Echo is described as a hands-free speaker that an individual manage using one's voice. The Echo is linked to the Alexa Voice Service to offer substantial benefits instantly. Alexa has been described as the brain behind the Echo.

Bringing the smart home to life

Many gadgets exist at home, but while trying to use them for improved home automation, they only lead to confusion. Amazon Echo has therefore comes in to improve such conditions. It has taken a lead as a smart home hub in the current market. The Echo renders it easy to manage smart home device compared to others. Not to mention that the Echo could be finicky about statements at times.

The Echo connectivity ability to IFTTT provides extra skills to its speaker. IFTTT employs the simplest rules to link to the apps, services, and other devices. IFTTT gives proper configuration and provides the meaningful programs in tech.

The Echo speaker can be connected to smart devices and tablets through the Bluetooth. It gives quality sounds. Additionally, it provides a rock-solid signal when employed as a primary audio source rather than the smart gadgets.

What Is Alexa?

Alexa is a voice command technology that was initially launched in 2010 to expand the Doppler technology portfolio. It was expected to perform beyond the initial Kindle e-reader. It operates on the Amazon Web services and uses Wi-Fi Internet connection. For it to function, it must recognize the words "wake word." However if one have such a name, it can be changed via the free Amazon Alexa application. From the time it was launched, it has included more than 100 new features and expertise. Not surprisingly, the Amazon Echo plays music and one cannot yet manage the Apple Music via the Alexa app. Though, it can be used with the Apple Music since it has a Bluetooth speaker. Moreover, Alexa has several works in the Amazon Echo device. Use the commands to control the Echo. For example, say, Alexa, next song.

Notably, the Alexa homepage gives a clear outline of the features to use to discover suitable add-ons. The search feature keeps being improved and they have input a device to add more apps. It only needs command, "Alexa enable" then give your skill which is computable.

Precisely, every person would love to know upgraded features on their device. First, one has to subscribe to an emailing list to read about the new features. You may also search for them on the internet since they are readily available. When you got the Echo, you only need to ask, "What new features are available?"

Notably, when a new feature is added, there is a program for explanation put into the Alexa. This allows it to provide the exact details required via a question. The manuals guide on the effective use of the new talents in the device.

How to access the Amazon Echo from the Web

The Echo from an original iOS helps to access most of the Echo settings quickly. Just browse to http://echo.amazon.com. This will help you access the significant details on how to use the Echo.

There is a way one could link the family with once prime accounts. It is accessed through the echo.amazon.com. What one need to do is to find the settings menu. Then, scroll down the page to set up your expectations. It is more helpful if Prime members link people than if they are not members. For such shared membership, one must download the Echo app to their smart devices and have a consensus on how to connect.

Overview of echo operation

Amazon Echo listens to every speech to monitor the wake word to be uttered. Notably, it has a remote control which is manually and voice-operated to pick the wake word. Additionally, device microphones could be manipulated by pressing the buttons provided. For instance, the mute button closes the voice processing circuit. Note that the Echo needs Wi-Fi connectivity for operation. Thus, the Amazon Web Services determines the Echo's voice recognition potential. Such services are generated from the Yap, Evi, and IVONA. The Echo device operates correctly with an excellent internet connection due to its capacity to lower the processing time. Therefore, this incorporates minimal communication times, streamable replies and the widely distributed operation endpoints.

Functions of Amazon Echo

Significantly, this device can be utilized in voice interaction, setting off alarms, performing music playback, producing to-do lists, playing audio books and streaming podcasts. Additionally, it provides information on the weather conditions and the traffic. It can be automated, and the content given is sourced from both the local and the global stations.

It has advanced considerably and has gained a lot of attraction from the consumers. This has resulted due to its widespread to many users across the United States. In fact, the Alexa voice operation is available for incorporation into other devices. Other firms are thus motivated to start connecting to the instrument. It has a built-in application to stream music work for Pandora and Spotify. Echo would undoubtedly help to access Wikipedia articles and music such as Apple music can be streamed. It could be either via a phone or a tablet.

Meaningfully, Amazon Echo will reply to the queries on the items identified on the Google calendar. Its integration with various devices has enhanced efficiency in servicing. Indeed, this forms a transparent platform for one to make the right decisions.

Moreover, the Amazon Echo can obtain expertise created with the Alexa Skills Kit adding its capabilities to function.

Interestingly, the abilities compose of the skill to play music, respond to general questions, set up an alarm, order foods or any other service delivery. The continuous addition of the skills boosts the abilities this device can offer.

Notably, the Alexa Skills Kit composes of several documentations on and actions from the codes used to add any expertise to the Alexa.

The Smart Home Skill API is newly developed software that teaches the way to manage lighting and thermostat gadgets. The codes operate in the cloud since there is no sure thing running on the user device. Module training or manuals are readily available for the way to use the emerged and the existing applications.

Building and managing a smart home

Significantly, the Echo integrates with many household automation centers like Wink, Insteon, and the SmartThings. It is possible to make one's home smart even if they reside in rental houses. Simply get a list of such smart devices online. This will guide on the up to date services. It caters for everybody despite the fact that some rent houses are not very spacious, and the tenants have little money to spend. This device can manage the lights, the apartment temperatures and assist in turning on one's video recording using natural commands. This is an efficient service that Amazon Echo offers to the users despite their financial constraints.

It is crucial to note that the voice commands vary based on the devices one need to use and the kind of names you give to the Alexa app. Some of the examples have been highlighted in this book. Others may include, Alexa, switch on the bedroom lights. Modify each of the commands according to what you expect to know. There are plenty of activities that one can do with the home automation devices.

Get the Weather (and Other meaningful Details)

Be sure to find weather information from different places and time. The details provided help one to schedule carefully on the relevant plans of the daily routines. The necessary information is a clear guide to what you want to avoid inconveniences. For weather information, you may try, Alexa, what is the weather? Or what is the time?

Additionally, one may set up news updates and other content in the Alexa app to help in enquiring more details.

Listen to a publication or a book

Unsurprisingly, the Amazon Echo incorporates with the Audible firm. Here, one can navigate places using Echo with easy voice commands and search for your location. It could be good to use a sleep timer whenever you want to relax while listening to the Echo. Find more Audible commands online depending on what you want.

Additionally, Amazon Kindle books can be heard via voice synthesis since some may lack audio companions. For example, Alexa read Funny Kids. It will simply read it for you.

When it comes to articles found on the Wikipedia, modify the commands to fit the question at hand. For example, Alexa read Wikipedia Trailer. One can still enquire how many more of the topics one has requested in advance.

Interestingly, Alexa can trace your childhood recorded actions. This could be either time spent with your family members and when you listened to cookbooks and articles before sleeping.

Other general uses

Echo replies to an enormous amount of fun Easter eggs which one can try out different phrase severally. You may use the Reddit link to obtain a massive list of comments. Whenever you require a full self-serving answer, ask Alexa to "tell a story." It has been very useful with simple math, and one can request Alexa to add values. The calculation could go pretty well.

Interestingly, the Amazon Echo can quickly repeat a solution in case it have answered in a very quick manner that one cannot understand. Just utter the words "Alexa, could you repeat that? It will always reply to your queries.

On the other hand, the Amazon Echo deals with basic date calculations. Here, one can enquire the number of days between certain dates. For instance, you may inquire, "how many days until November 21st." It understands the extended holidays like Christmas and Halloween. It is impossible to say exactly the number of days until the Super Bowl

Seemingly, there exists actual humankind who can respond to questions regarding your Amazon Echo. Input your number on the http://echo.amazon.com/#help/call and there will be a person to call back. Amazon has demonstrated that most people cue to get the required help. Thus, this gives a sound outline of how you can attain maximum help from the Amazon Echo. Try to stay tuned.

Thus, the Amazon Echo provides an excellent glimpse of the bright future home. Whenever you hear people talk about the Alexa, do not get confused about what it is capable of doing. This book outlines the proper things one can obtain from the Amazon Echo. All you need is to learn how to integrate it with other devices to offer maximum support of what you need. Without any doubts, you will love it since the book gives various exact things the Echo gives and how to perfect its functioning. Always keep in mind that the Echo is a speaker that plays music in a variety of unusual ways that this book will discuss.

The Echo has advanced over time making the Amazon library great to offer various types of music. The playlist would suit each mood for a given moment. For instance, one can request Alexa to play Christmas music, Adele or meditation tracks. The current modifications have also enhanced the use of the Spotify, and if you pay for its access, you can get everything you require via the Amazon Echo. Here, one can request for genres, music or composers provided that one specifies "from Spotify." Follow the right commands to explore the benefits accrued to the incorporation of other devices to Amazon Echo.

Features

Voice system

The Amazon Echo produces natural, lifelike sounds from the speech unit selection technology. Increased speed precision is acquired from the sophisticated natural language processing algorithms made in the device text to speech motor.

Application updates

The updates are made occasionally based on the software that is released by the Amazon group. Therefore there is increased accuracy of the functionality of the Amazon Echo. The recent updates fix bugs that are crucial to enhancing satisfaction to the users. The process of installation of the emerging releases is on a gradual basis from one release to the other. It ensures complete installation of the entire advancements.

Additionally, the Amazon Echo has some advantages in that important enhancements could be created even without upgrading the functioning software model.

On the other hand, Amazon Echo hardware has a 4GB storage space and a significant processor. The Echo is connected via double-band Wi-Fi. It has a Bluetooth 4.0.

Input

Manipulations are made using the remote controls to control the Echo voice keenly. The user setup could be modified from one region to the other using an action switch on the apex of the unit. The upper half-inch of the system revolves to lower or raise the speaker sound. The Echo have to be charged to run because it does not possess an internal battery.

Apps that boost Amazon Echo functionality

Automatic app

There are useful apps which made the Amazon Echo entirely intuitive in serving as a digital assistance. Some of the killer's skills do not need additional hardware, but the integration with other software has made it work correctly. Those who own automatic smart vehicles are liable to link the account to the Alexa app and inquire anything on their automobile state. For example, it is possible to ask Automatic the distance one drove a previous day. Also, you will know the level of the cars fuel to plan ahead.

Capital One

This brings capabilities that allow the Alexa to pay bills. A moment the Capital One account is linked to the Alexa app, one can ask various updates, for instance, the card balance and the transactions statements. In the new future, this app will help transfer cash.

Domino's Pizza

This allows one to order pizza whenever needed. One can request quick delivery of such goodies. Only Open Dominos and seek an order. If one get impatient, confirm from Alexa the status of the decree.

IFTTT is responsible for the incorporated skills that help order pizza or any other foods with the voice. A Twitter account assists this. To start up, go to the Dominos.com and then log into your profile. Where there is a Pizza profile page, get logged in for Tweet ordering and link to the Twitter account. Here, there is the Domino's page where you get to see the previous orders where you save the one you want. When the Domino pizza is connected to the IFTTT recipe, one is notified through a direct message.

Jeopardy!

This comes as a surprise to everyone. Jeopardy skills allow users to play their favorite game shows. This flexes one's knowledge.

Quick Events

This allows tapping and swiping of activities using an intuitive scheduling ability. It makes it possible for the Alexa to remind one on a busy calendar schedule. Be sure to never miss an event with the presence of Amazon Echo.

NBC News

The NBC News skill put the latest talks within the voice reach, for instance, the elections. Such data include the primary results, trivia, and any breaking news. It thus delivers important updates to the users.

Spelling Bee

This is an exceptional skill that helps the Echo to repeat the fifth-grade level statements which the user need to spell loudly. The Echo knows the way words are spelt and acts as referee. It may be a fun endeavor to pass the time.

The Wayne Investigation

It helps one to select the desired game to adventure.

7-Minute Workout

It offers exercises that are scientifically proven rendering people to a healthier lifestyle within a short time. It gives the equation of how such actions are required to take place. Only connect this app to Alexa and command it to begin seven-minute workout. Then, the speaker takes control, but the skills allow some breaks. In future, the talent is said to be advanced to track the statistics and provide more challenging practices.

Uber

Ideally, the Echo searches you a car to reach your destination. Just enable the Uber skill to the Alexa app and sign in with the personal details. It's nice to include one's location to alert the driver on the place to pick you. The kind of ride one gets is safe and secure. Alexa will answer any questions regarding the distances moved and the designated stops. Indeed, you only need to add such content to the device.

You must agree to the terms and get the device situation. Probably, you can order using such word, Alexa, ask Uber.

The book has illustrated a good reason for one to purchase Amazon Echo at Amazon. You will love it so much and opt to own one in your house. Ensure proper naming of the Amazon Echo.

Play ball

There is a high number of individuals who consequently enquire on their favorite sports groups. Well, the Amazon Echo can get the information like the scores and the upcoming schedules for the sports teams. Only quote, Alexa, "show me my sports updates."

If you like to set up the favorite team in the use the menu settings and choose the Sports update.

Traffic alerts

The Alexa provides crucial details on the traffic issues to avoid lateness for once business. It helps avoid congestion. First, the device must know one's home and the workplace address. This is included in the app using settings menu where one goes to Accounts menu bar and chooses Traffic. It is possible to include any significance stop along the road. Importantly, every contact should have the street number, identity, the town, and the state. To add the address inquiry from Alexa, simply commute and the Alexa gives you the right traffic.

Music playback

Ideally, the Echo helps to play music through its speakers. It is fun, and it is very likely for an individual to like and use it frequently. Notably, the Echo is linked to the Amazon Prime account that allows one to access the purchased music and the free music in the free Prime music. Here are some of the options one would give to Alexa to listen to music:

- ✓ Alexa, play (music genre)
- ✓ Alexa, play (band)
- ✓ Alexa, play (song by name)
- ✓ Alexa, what's playing?

Significantly, you can listen to the music playlist you have created via the Prime music system. It is possible to give the songs the preferred names. Thus, the Echo offers lots of fun.

Inquiries, Conversions, and Trivia

The Amazon Echo is relevant in an office for a variety of reasons. If one has a lot of queries, the Alexa app can handle a broad range of topics and provide considerable solutions. In fact, she will either provide a direct response or it doesn't have such database;

she will reply using the solutions from the search results. This would just be like Siri.

Commands would be;

- ✓ Alexa, what's new today? (It could be weather issues of the workplace)
- ✓ Alexa, what is (calculation problem e.g. math issue)
- ✓ Alexa, what is (give the question about your office work, it could be word spellings)

Seemingly, Alexa device in the Echo is quite helpful in handy moments. It prevents doubts by giving the information required at a particular time. For example, it can tell how many employees an individual firm has.

It is possible to ask trivia questions from the Alexa which is very enjoyable. Well, it doesn't offer perfect answers but gives a good solution to problems. For instance, one may inquire about the speed of different birds, the size difference from various planets, famous people or any other specific questions.

Shopping and To-Do lists

Use this device to add items on the shopping list. Most people get to realize that they are out of something when they have already started their activities. There are very simple commands which help to include certain things on one's list. There is no need to wary since the records readily transcribed for you. For example

- Alexa, add (name of the item)
- Alexa read my to-do list

Seemingly, each item you include on the Alexa app is stored. Thus, it is necessary for referencing once you run errands. The procedure is instantaneous making a quick entry to the list on the app.

How to use Amazon Echo

First, generate the desired wake word. It could be Amazon or Echo. It is critical to note that if your name is Alexa, it is good to modify the "wake" word which alerts the device to begin listening.

One must be prepared to handle growing list of the third party gadgets which are incorporated with the modern Alexa. In fact, there are a lot of details that one can request digital voice to perform.

Now you can go to app settings and choose the profile to use. Here several guidelines are issued to help you use the app. You will need to enter account details, and help menu bar will guide you on the setup. Thus, it is easier for one to switch back and forth between the various accounts. Only say "Alexa, switch profile" and it will shift to the account. Since there are a significant number of skills developed for Alexa, you can try any and she will do what is expected. These could be games, quizzes or fitness routines. In case you feel that whatever you have asked Alexa is irrelevant, go to the settings and search for history. Select particular recordings and delete them. Whenever you need to erase anyything, check in the www.amazon.com/myx, select "your Devices" click on the manage voice recordings and then delete.

Managing the Amazon Echo multiple Amazon accounts

Ask Alexa for the profile that you want to use and switch to the desired profile as you have named. Every profile will acknowledge the gadgets linked to the Amazon Echo. For pattern groups, set up the group first and then give names to each of the profile. Make sure that devices are controlled for each of the group.

Various places where one can use the Amazon Echo

In the kitchen

Alexa could be of significant use in the kitchen because your hands are always busy preparing foods or doing other kitchen chores. It is possible to set a timer for the meals one is preparing. There even are some apps on the Echo that have a certain recipes which could guide you in making the meals. She, Alexa, can use a Bartender and Drink Boy apps to teach you on how to prepare a cocktail. Others, like the wineMate can provide recommendations on whatever you are cooking. If you miss some point, you can ask Alexa "Can you repeat that?

Interestingly, when you run out off something to prepare the dinner, it is possible to request Alexa to include that item on your shopping list. Reordering items like serviettes is possible, say, "Alexa, reorder serviettes." In case you change your mind, just cancel the order immediately.

Notably, Echo is a great device that makes Alexa an excellent kitchen assistant. Well, she perfectly does a lot of work. Alexa can help in measurements and units conversion for the foods being prepared. Set timers to prevent food from burning.

Questions like, Alexa, how many tablespoons are in the butter stick. Also, Alexa set a 20 minutes timer. Additionally, Alexa can read Dump Dinners which composes several recipes at the required time. It lowers the time needed to cook while learning. Just listen to Alexa while in the kitchen.

In the bedroom

Alexa is available to have a smooth running of the daily activities. In the bedroom, one can request it to set an alarm. You can wake up call daily as required. It is simple to edit the set alarms and modify the tones to avoid disturbances during the sleeping time. It is crucial to delete an alarm or utter the words "Alexa, cancel signal for" and then choose the day. Select the exact times you want to set the alarm. If you forget the time you are expected to wake up do not hesitate to enquire, Alexa, "when is my alarm?"

Consequently, one may ask the Alexa to switch off the lights and lower thermostat. A significant thing is to have the required compatible smart home apps. Alexa will also guide on the traffic, weather and information briefings. Indeed, the Echo will always show Google calendar where one can add favorite events. When late for work, you can easily call a taxi.

In the living room

Entertainment is a primary aspect in the living room. Thus Amazon Echo has considered this factor and can allow anyone to listen to music. One can enjoy everything from the Amazon Music, simply by requesting Alexa to play it. The only thing you need to possess is the Amazon Prime account. Nevertheless, those having Spotify Premium account may sync music using the Amazon gadgets. Others could be the TuneIn, iHeartRadio, and Pandora that will run the tunes, internet radio and podcasts upon request.

It is possible to purchase Taylor Swift MP3s and save into your Amazon Music Library. It keeps up to 250 songs without charges which one can listen. To upload them, use the Amazon Music desktop.

For the audio books addicts, it is possible to apply paired Audible account. It 's nice to identify the list of eBooks compatible with

Alexa so that it can read it for you. Be sure to experience a lot of fun as long as you have purchased Amazon Echo.

It is guaranteed to get the kind of movie one wants to continue having fun with Alexa. This device has lots of jokes and Easter eggs with great replies to all the queries. Enable such skills to adventure on their endless entertainment.

Tips to maximally benefit from the Amazon Echo

Always feel in the correct position while trying to use Amazon Echo. That said, there is a quest to study smoothly on the effective ways to use Amazon Echo. Importantly, this guide teaches you the proper ways to utilize Echo machine.

Device only responds when it hears the wake word. The mute button helps to keep the radio silent. One should press it until it shows a red ring and Alexa will keep quiet unless you push it again. This could be utilized during meetings to avoid noise disruption.

Ways to add a software update to the Amazon Echo

The Echo possesses a CPU at its core which helps to run the software like any other digital gadgets. Therefore, it means there is a need to upgrade. Importantly, Alexa looks for updates every night. Other updates could be done after muting the device for at least 30 minutes.

Amazon echo connectivity

The Echo is linked to the IFTT where one can automate everything. It can also save some money for you. A great deal of the possibilities could be checked on the IFTT tag page.

Sometimes one can request for whole recipes using secure command after installing the set ups. This could assist you to get the easiest recipe to follow. For example, Alexa, "trigger Party Time." Furthermore, one can get the entire shopping list via email in the Echo Alexa app which renders the feature very useful. On the other hand, not all people need to use a voice command. You may have the IFTT log the music you want to hear in a spreadsheet. It eases the playlist. The use of IFTTT gives endless capabilities, and one does not have to employ the pre-written recipes. You may customize your recipes and request Alexa to activate them. In fact, this will amaze you as it will still be useful.

The trick is to ask what you want. The Amazon Echo possesses seven microphones and beams forming techniques to hear the voices across an entire building. It does so even while there is music going on due to the enhanced noise cancellation. The device's speaker is expertly tuned to occupy any space with 360 degrees immersive sound. Just vocalize the wake word "Alexa" and the Echo will respond instantly. In circumstances with more than one Echo, make sure to set a distinct wake word for each. Alexa gets smarter every day. The more it is utilized, the higher it adapts to one's trends of speech, personal choices, and vocabularies.

It is a Bluetooth enabled and capable of streaming other known music works such as iTunes from the once smart device. Tunings allow it to give crisp vocals that produce lively bass. In fact, Echo can hear one asking queries from all directions. Since the Echo is connected continuously, briefings are delivered automatically.

Recent updates included new skills for the local search from various third party reinforcers. Simply discover lots more with the Amazon Echo. The ratings and reviews for the added skills are readily available on the Alexa App so that you can learn them gradually. This way, one learns what other users have to say about different expertise on the Alexa App.

There are starter's kits on the various devices that work with the Amazon Echo. They make setup incredibly easy. Free Alexa App on the Fire OS, Android devices, and the desktop browsers help one to set and control Amazon Echo efficiently. Smart devices like WeMo and Philips have the possibility to search for Kindle books to read and Audible libraries to use. One can see shopping and to-do lists while browsing.

Timers are controlled quickly and require customs tones put. In fact, the Alexa App helps to discover a lot of things and allows installation of the third party expertise.

controlling your smart things

Alexa can be used to manage cloud-based gadgets, though they must first be set up via Connect Home Alexa section. Choose the option "to discover devices" and search for compatible devices on similar Wi-Fi connection.

Ideally, Amazon.com is the greatest online store in the universe. Thus, learning the Amazon Echo to carry out a good job in helping make shopping lists and spend the money. Alexa can include to-do shopping list. Conversely, the records can be accessed from the Amazon website. Purchase of products is through the Amazon.com, and the items are easily obtained using Alexa app. One can provide a shipping address on the file. In fact, it is an amazing process as one orders the item needed. Alexa searches through the history of your previous similar orders. Additionally, one gets suggestions on the things to buy since it prompts one before usage.

How to set up the Amazon Echo

- Prior starting to use Amazon, plug the device into a power outlet
- Download Alexa app from the App store. Note that the 1st and 2nd generations tablets are incompatible with the Alexa application.
- After plugging in Amazon Echo to power, the indicator ring at the upper part flashes blue and then studs to a rotating orange color. Thus, this is a clear indication that device is ready to configure with one's WI-Fi connections. In case, the ring changes to purple it won't work unless you keep holding the action stud until you get the orange color again.
- Open the Alexa app and turn on the Echo. Avoid lousy Alexa by setting the speaker's top to reduce the volume.
- You will note that the Echo links to the double-band Wi-Fi that is of considerable standards. Amazon Echo does not connect to ad-hoc networks.
- In the Alexa app, open the left navigation toolbar and click the settings button.
- Choose the device and then update the Wi-Fi. Select a new device to your account wherever necessary. Consider restarting the Echo device if it does not connect to the Wi-Fi connectivity. When such problem persists, reset its factory settings and try again.

Use the Action button on the Echo device to connect it to the mobile machine. Available Wi-Fi networks pop in the app. At times, the app could require that you connect your device manually to the Amazon Echo gadget via the Wi-Fi settings.

Simply choose the Wi-Fi connection and enter the password. In case, your Wi-Fi connection is invisible, scroll down and select an option that adds a network or try rescanning connections available.

Use the MAC address to include your Alexa device to the routers approved gadgets lists.

It could also be crucial to save the Wi-Fi password to the Amazon though this is optional. It would thus be remembered whenever you want to reconnect to the internet.

Another option is to link to the public network which just requires you to enter the necessary details. Such information is not likely to be saved in the Amazon.

- Finally, select connect and a notification pops on the app where it is possible to start using the Alexa. The next thing is to talk to Alexa by using the Echo machine.
- Use "wake word" and begin speaking naturally to Alexa. By default, the Echo recognizes "Alexa", though you can modify the wake word through the app settings. Go to the option Wake word under the settings menu.

It is advisable to note that like any other machine you need to take extra care in the way you handle physical manipulations. Importantly, the volume ring, the microphone switch, and the action knob should be well maintained. This should work efficiently to generate the required information. The microphone stud turns off the listening gadget whereas the action button summons Alexa and initiates a manual setup if held. The volume ring helps adjust the volume levels.

Summoning Alexa: Wherever you mention the word Alexa, you will get the top spinning blue. It will orient itself with the brightest point towards the sound source. Then, one may inquire anything he/she needs. The situation has proved the use of advanced technology to do the significant things in life.

How to customize your Echo experience

Notably, this book has grouped most significant usage of the Amazon Echo. However, when one is chatting with Alexa, there could be the will to adjust settings to customize one's experience. For instance, when you ask for "what's the current news." It could be done through;

- Open the Alexa app and navigate the settings on the primary menu
- Scroll down to the "Account" section. Here, there are variety of options, like music where one can connect the Echo to the Pandora account. Others could be Sports updates, and one can choose their favorite teams. Additionally, you will find the traffic briefings, calendar schedules and the places for connecting linked home brands.

One may set what they wish to hear or get and make it the default options. For example, choose whether to receive flash briefings from the BBC or NPR news and make it the default station. Simply navigate the Setting section, go to the "Account," then select "Flash Briefing" and then switch to the selected position. Similarly, you may wish to leave both on and receive back to back news updates. Such steps will have your Echo properly set and ready for you to use it correctly.

You may play around with the device and invite a friend who would surprisingly ask entirely new questions about her. If there is a pressing question concerning the Amazon Echo and how to automate it for home use, post the query to the Amazon Echo support team to get solutions.

The Echo basics you should always remember

- ✓ The Amazon Echo is not merely a speaker. It has a firmly controlled voice by the Alexa. The Alexa is an artificially intelligent device that renders the Echo a multipurpose tool. It is an outstanding individual assistant in daily works. If you already possess an Echo, or you consider purchasing a new one, this book is an appropriate guide on how to make the best out of the Amazon Echo tech.
- ✓ The Echo is attached to a power plug which makes it work efficiently. Initially, the device had a remote control that possessed a microphone though recently Echo shipments have inadequate accessory.
- ✓ Importantly, one can listen carefully to people across borders provided that you have raised your voice.
- ✓ The Echo set up is quite easy. Simply put in the speaker, download and open the Amazon Alexa application. Follow the prompts.
- ✓ The Echo uses the Bluetooth speaker to link the home Wi-Fi connection to produce voice summons over the Internet.
- ✓ The Echo's accompanying applications need to sign in through the Amazon user account. Thus, this connects the speaker to most voices recorded over the years. For instance, one can obtain Kindle e-books bought a long time ago.
- ✓ The app can be used to modify Echo's settings. The two distinct choices one must play with are the "wake word" which the speaker listens. The Echo then interprets the information following. Alexa is the default wake word. Using it is fun.

- ✓ The other key choice for the Echo is how to enable awake sound. The confirmation tone plays after the speaker listens to the wake word.
- ✓ Whether the wake word is turned on or off, a light blue ring shows on the top after the wake word detection.
- ✓ Both the light ring and the wake voice act as a warning. Amazon Echo listens to the voices all the time. Use the microphone off option to turn device deaf though Echo is constantly on alert. When it gets some views, it records the audio instantly. The recording gets streamed over the Internet to Amazon. Here, they are continuously analyzed and converted into commands on the Echo. This is similar to the way Apple, Google, and Microsoft voice assistants.
- ✓ It is possible to obtain the history of the entire summons via the Amazon Alexa application. If you like, they can be removed. The more the Echo listens, the more efficient it becomes at interpreting voice trends and idiosyncrasies.
- ✓ There have been voice-activated personal assistant techniques prior the Amazon Echo, though the Amazon Echo is highly competitive. It has many online retailers who subscribe to the Amazon Prime where they make several purchases.
- ✓ The Echo can associate device to the Google Calendar which enables easy tracking of any appointments. Once you connect the speaker to the Google account, only ask Alexa and the speaker will give you details of what you expect.
- ✓ Ideally, the Echo searches for content on local business via Yelp. The search results provide an update of the briefings. The Echo do not require outside help for each of the personal assistant attribute. The timers and the alarms get controlled by the gadget and its app. For example, Echo's to-do list attributes allow the users to retrieve reminders at their speaker. This gives remainders for the activities done.

- ✓ The traffic update feature is extremely useful. Any other information is obtained quickly from the device. It can give other users like the Android and the iOS.
- ✓ Notably, Echo operates well with the natural thinking of the people and the way they behave. You can listen to the slow or loud speech or get any spur for a device action. The Echo device has partnered with TuneIn to allow streaming of local radio stations. Only, tell it to play your favorite station by giving its call letters. Live streaming is possible, and one can play podcasts through requests of shows by names.
- ✓ At times, device gets so much of Internet content at its disposal which confuses it. For instance, saying Alexa, play Serial, can give different things. It could end up giving a song for " Cereal Killer" instead of giving Serial podcast. Thus, make it a habit of giving a clear outline of what you want to get for it to function well. The conversation would be of high quality since its tone would clearly comprehend on the users' questions. Also, it can find information on Wikipedia and other sources. It can, therefore, give replies to the multiple questions though the results could be spotty. A good example is where the Echo can tell the best movie during an individual Awards ceremony in 1980 but cannot give details of the presidential candidate winner.
- ✓ Importantly, device searches through once previous orders and issues options. It recites the goods information aloud, includes price and inquires whether one desire to place another order. In fact, this just works perfectly. For the Amazon, this kind of shopping is better since customers will not have to keep comparing products with the competitors' brands. Thus, this lowers competition levels.

Indeed, it is possible to manipulate nearly everything that the Echo does with your voice that makes it more appealing.

Some say that the Amazon Prime Music limits one to what to listen to, but Alexa allows connectivity to various devices to achieve the users' satisfactory levels. Despite several controversies, the Echo remains a suitable device via the multiple skills developed inside. There is a high likelihood of creation of more competencies in the Alexa appliance. More new tools give Alexa many hardware specific apps making device more intelligent and efficient in service delivery.

The Alexa platform integration with various third parties and replenishment applications such as Dash gives a real ambitious home via the Internet. The Echo was responsible for the rise of the Alexa ecosystem and has led to several supporting operations which help to control other home devices. Significantly, Alexa functions beyond critical since it boosts the convenience degrees that it becomes difficult to live without it. Opening up the ecosystem to integrate with several devices expands its capability. Other gadgets like Google have tried to use the skills since it has proved to attract massive numbers. However, Amazon Echo gave the satisfaction prior the others making it remain competitive in its usage. The Amazon Echo continues to own great advantage.

The Echo device has supplied consumers with a great deal of their expectations that is too relevant to their lives.

The principal advantages of Echo over its competitors

Echo has more data

The Amazon industry has ensured maximum data availability to allow interactions with several users. The firm has used data to consistently boost learning algorithms of the machine and natural language producing ability behind Alexa. Alexa could be a great help in gathering information from various customers during the interaction improving its functionality.

AWS Backbone

Codes operating on the Alexa system are in the Amazon Web Services cloud system. Very few codes operate the Echo gadget. The phenomenon has played a significant role as it permits Amazon to provide updates to the Alexa device frequently in minutes. This makes it easy and quick to advance and add other skills to allow Alexa communicate with various services, like Spotify. One could say that such capabilities could take the responsibilities of smart apps in the Alexa ecosystem. Its dominance in the market and the flexible AWS infrastructure leaves a significant challenge to its rivals to compete with the Echo device.

Innovations

Ideally, Amazon has the will to keep its technology on Echo related capabilities. Additionally, Amazon Company has invested heavily to allow maintenance of the Echo ability. Such funds aim at improve designs used in developing software that could ease functionality of the Echo device. Such actions will improve the initial Alexa ability and lead to substantial long-term benefits compared to other players. The Alexa experience is fun and quite innovative because of the diverse features.

Availability

Apparently, Amazon Echo is readily available for everyone to purchase. It has moved from the past where an invitation was needed to buy Echo. The pricing is relatively considerable for both the retail buyers and the Prime members. There is always a reason behind the need to purchase Amazon Echo.

Amazon Echo Pros

The list below highlights all the accrued benefits one would get if they owned an Amazon Echo.

- It entertains a lot through music. It is fun to play songs using verbal commands
- It is easy to obtain music from one's Amazon library, Pandora, etc. The Prime members acquire the Prime songs. It is possible to upload music to the Amazon Prime using Echo.
- It gives different information for various fields. It integrates appropriately with other devices to satisfy the user's needs.
- Echo can listen to an individual from a long distance even under noisy condition. It has sound and fast speech recognition when it understands the commands issued.
- The Amazon Echo has many features which are continuously being added. More features are expected in future for advanced operations.
- The Echo has a remote control which is like that of the Fire TV.
- One may request for Echo fun facts
- It also has many jokes and games which serve to entertain kids
- The current price is considerable despite the fact that it is not just a speaker.

Amazon Echo Cons

Like any other device, Echo has its limitations;

- First, device has the potential to upgrade its audio. It has good sound but could be better.
- Inability to search within a particular playlist
- It needs effective intervention via an alternative user interface for the successful buying of the digital media. Other times, the Amazon Echo shows the hit or miss outcomes for regular questions that any consumer would expect high-quality solutions. Apparently, the device location set to Seattle by default and requires manual modifications n regions around the United States. Consequently, there could be undesirable or misconception solutions for queries on weather or time scheduling. Time must thus be modified regularly to cover the hour differences and report the right time. However, this would challenge the accuracy of information received by the Amazon servers. The chrome request maker would assist in finishing task successfully. Also, the fact that Echo is only available in English would limit number of users.

Suggested Improvements for the Amazon Echo

There are many suggestions on the way to develop Amazon Echo to a much better device. Such recommendations are based on the user's feedback.

- First, the audio and bass should be better and deep respectively.
- There should be multiple music choices to outdo the competitors. This will avoid music getting quite repetitive first.
- Device strategies that would permit the pairing of Echo for use as stereo speaker for home theaters gadgets.
- Improvise an optional battery pack for external purposes.
- More support through the Skype and awareness creation.
- Ability to predict the next happening to Amazon or Prime Videos
- Make Echo an excellent speaker before adding any other features.
- I would also recommend Echo for individuals with a liking for gadgets since it is a great conversational piece. The new version features must, therefore, turn Echo a better audio than any other competitor.

Echo and the future

The Amazon Echo has just been used for some time. Thus, it is targeting to be more active. The app development would be a significant drive to satisfactory functioning of the Echo. More investments are required to develop such platforms to promote the number of skills. The stock-quoting capabilities must also be maintained on a long-term basis. The Amazon industry must also work towards the challenges that have been posted due to its usage. This will reduce too many frustrations from their clients. Similarly, this will improve the level of categorization of the skills available to avoid disorganization. It has to maintain the good start.

Amazon echo privacy concerns

Privacy matters have become a great concern for any device nowadays. Thus, the Amazon Echo must ensure the safety and privacy of conversations. Most likely, people would like the identity of the individual present at home hidden for certain reasons. Audible cues like the footsteps or TV programming could reveal the identity. Significantly, the Amazon Echo has catered for this issue via streaming recordings from each consumer's house when the wake word activates it. Nevertheless, it is capable of streaming the audios at all times and listens to the user's word for detection.

The Amazon Echo employs the past voice recordings sent to the cloud by the users to promote replies to future inquiries by consumers. The user can remove audios which are closely associated with his/her account to address privacy matters. However, it is believed that such acts would degrade the user's expertise for applying the voice search. You may visit the Amazon customer service or use a setting to manage Device page to delete any information.

The Echo employs an address created in the Alexa companion applications whenever it requires location. Thus, it uses location details to offer location-based operations and keeps the information to give voice functions. For instance, the Amazon Echo sound activities use strategic location to reply to the user's request for nearby services. Additionally, it needs the user's location to generate mapping related requests and boost the Maps experience. The entire details gathered are entities to the Amazon.com Privacy Notice.

In fact, Amazon maintains the user digital recordings uttered after the "wake-up word." The recording should comply with the laws enforced and the governing policies. Warrants and subpoenas accompany the device. There are clear indications of the illegal acts the device gives.

Conclusion

Back in the day, many homes used radios to listen to music, news, and sports, etc. It wastes lots of time due to waiting for information you desire. Apparently, the Amazon Echo has simplified delivery of the needed information. It can act as a home built app. Just ask Alexa to give you any detail you require. Exploring the Amazon Echo is essential. It will give the flash briefings and the latest news. The Echo is excellent at comprehending context of what you need. The Echo serves as a perfect personal shopper since it is quite easy to place orders from the Amazon Prime whenever logged to Alexa app using the Amazon account. Goods are delivered right to your doorstep once using the Echo. If you like advancing with new technology, buy the Amazon Echo and encounter Alexa's benefits. Alexa is growing smarter daily since Amazon values innovation for their Echo product. Therefore, it will provide adequate service throughout one's activities. The Amazon Echo is a significant device that has led to the advancement of the activities described in this book.

Thank you once again for purchasing this user guide. I hope you have learned so much about the Amazon Echo.

www.ingramcontent.com/pod-product-compliance
Lightning Source LLC
Chambersburg PA
CBHW071830200526
45169CB00018B/1301